5
minute
OCEAN STORIES

WRITTEN BY
Gabby Dawnay

ILLUSTRATED BY
Mona K

MAGIC CAT PUBLISHING

In this book you'll find nine ocean
stories to read aloud, each one just 5 minutes long.

Each one is a wonder of the deep blue sea...
waiting to be discovered by you!

At the end of each story, explore an informative
'ALL ABOUT' page with a grown-up.

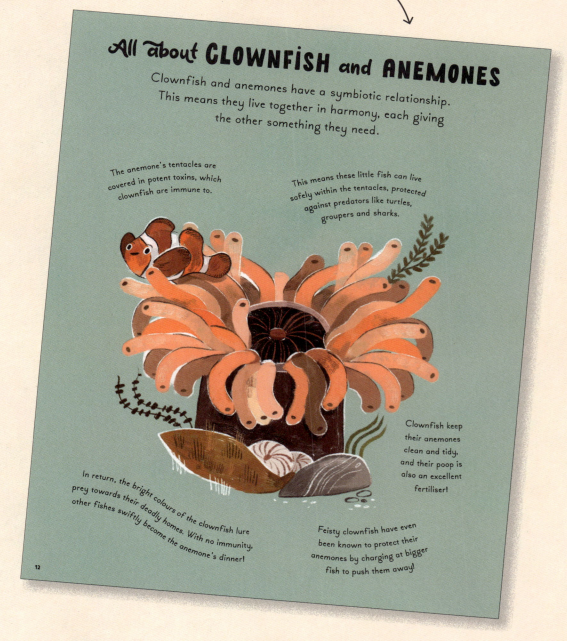

All about CLOWNFISH and ANEMONES

Clownfish and anemones have a symbiotic relationship.
This means they live together in harmony, each giving
the other something they need.

The anemone's tentacles are covered in potent toxins, which clownfish are immune to.

This means these little fish can live safely within the tentacles, protected against predators like turtles, groupers and sharks.

Clownfish keep their anemones clean and tidy, and their poop is also an excellent fertiliser!

In return, the bright colours of the clownfish lure prey towards their deadly homes. With no immunity, other fishes swiftly become the anemone's dinner!

Feisty clownfish have even been known to protect their anemones by charging at bigger fish to push them away!

12

Which **5 minute** OCEAN STORY
will you read today?

Out in the ocean of ultramarine
there's a world full of wonders that waits to be seen...

Every day there are wonderful treasures to reach
in the sparkling bay of a warm sandy beach...

From a rainbow of reef to the song of a whale –
there are sharks full of teeth and an eel with a tale!

Creatures that glow in the dark of the deep
or an octopus garden where jellyfish sleep...

In waters as salty as sea turtle tears
there are seahorses dancing away through the years...

Penguins that cuddle together and squeeze
on a platform of ice in the Antarctic freeze...

A sea full of magic for every child,
from a whale full of song to a storm full of wild!

Every flickering fish, every shell on the shore...
What a wonderful ocean of life to explore!

In this beautiful world,
every voice has a song.

HEAR THEIR STORIES UNFOLD –
ONLY 5 MINUTES LONG!

The CLOWNFISH, the ANEMONE and the BIG CORAL REEF

Out in the ocean
and under the sea
a little fish lives
in a tentacled tree...

She swims in the currents
and rests in the calm,
while the tender anemone
keeps her from harm.

This ocean in motion
is fully alive
with the humming of fishes
that flicker and dive...

7

In a reef like a rainbow
of tapestry, spread
in a patchwork of patterns
across the sea bed...

The water is clear and
the life here is teeming.
Content in her home
little clownfish is dreaming...

"Today I will leave
for it's time to explore
every coral and crevice
from shipwreck to shore!"

"I wonder what wonders await..." she begins,
as she dashes away with a flick of her fins.

Past seahorses, sponges,
a lobster and clam,
she swishes her way
through the sea-traffic jam.

"My wonderful city is ever-so-busy,
but all of this traffic is making me dizzy!"

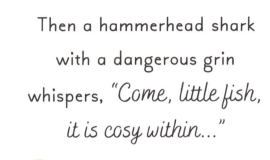

Then a hammerhead shark
with a dangerous grin
whispers, *"Come, little fish,
it is cosy within..."*

"Oh no!", gulps the clownfish,
and tries to keep steady,
*"I need to go home
for my supper is ready!"*

Next from the shadows
emerges a grouper,
"Delicious!" he says,
"Are you supper? How super!"

The clownfish replies, *"I am terribly small,*
I am hardly a meal — I'm a mouthful, that's all!"

With a flicketty-skip she is off and away
and then, *"Sorry!"* she hollers, *"No supper today!"*

She dashes through seaweed
and splattering foam,
she has almost arrived
at her tentacled home...

When...

"Hello," says a voice,
"You look lost, I can tell.
Shall I give you a ride
on my sea turtle shell?"

"No little fish!"
cries Anemone, "DASH!"
And the clownfish is safe
with a swish and a splash.

As the predator turtle
swims quickly away,
the clownfish calls out,
"Are you going? Do stay!"

"My tentacled tree
is the best place to be –
for the tentacles sting...
but they never sting me!"

"What an adventure!"
the little fish beams.
Then she brushes her home
'til each tentacle gleams.

All about CLOWNFISH and ANEMONES

Clownfish and anemones have a symbiotic relationship. This means they live together in harmony, each giving the other something they need.

The anemone's tentacles are covered in potent toxins, which clownfish are immune to.

So these little fish live safely within the tentacles, protected against predators like turtles, groupers and sharks.

Clownfish keep their anemones clean and tidy, and their poop is also an excellent fertiliser!

In return, the bright colours of the clownfish lure prey towards their deadly homes. With no immunity, other fishes swiftly become the anemone's dinner!

Feisty clownfish have even been known to protect their anemones by charging at bigger fish to push them away!

The water is waiting —
it's time to explore
every coral and crevice
from ocean to shore...

The GREAT WHITE SHARK'S
Tooth Machine

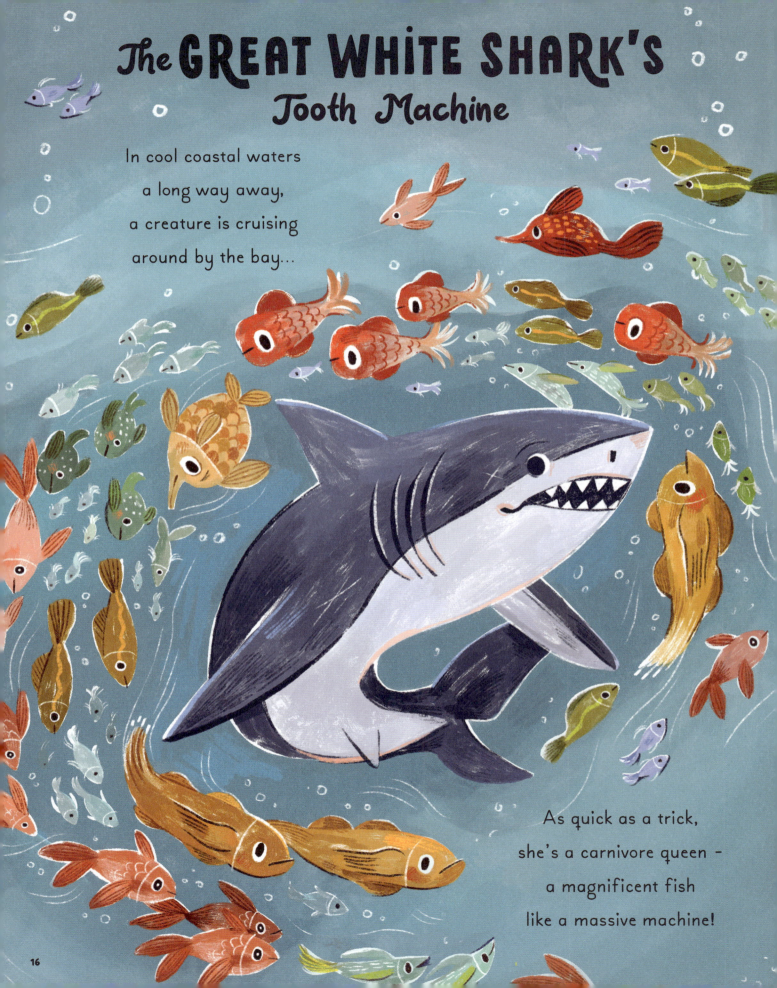

In cool coastal waters
a long way away,
a creature is cruising
around by the bay...

As quick as a trick,
she's a carnivore queen –
a magnificent fish
like a massive machine!

A bullet-shaped body
that's built to move fast,
see the smooth operator
accelerate past!

"I'm a fish with a swish and I cut quite a dash
as I sweep through the deep, racing past in a flash..."

"I am slick as a super-car –
what a design!
I am queen of the ocean –
the ocean is MINE..."

So silent and deadly, she slips through the waves
sending multiple fishes to watery graves...

She glides like a vision
of flickering night,
with a smile that reveals
lots of sparkling white...

Three hundred triangular
razor-sharp teeth
of a premium predator
lurking beneath…

*"I'm a winner at dinner
and everyone knows
every tooth is terrific —
just look at the rows!"*

"My mouth likes to munch
breakfast, dinner and lunch,
and a fillet of fish in between...
But whatever your theory
I'm really not scary —
I much prefer ocean cuisine!"

Jaws like a tool-kit of multiple knives,
the sharks' teeth keep growing
throughout all their lives...

"I lose lots of teeth
when I bite on a snack,
so my new ones are ready
to quickly grow back!"

"My fixtures are flexible,
that is the aim —
see, my gnashers are set
in a cartilage frame…"

"For I am a hunter,
and terribly clever —
I need lots of teeth
so I'll grow them forever!"

"A constant supply —
a conveyor-belt feed
that will keep my mouth full
of the teeth that I need!"

A shark
with a smile
full of sparkling white...

...who glides like a vision
of flickering night!

Now look on the floor
of the ocean beneath...
see those gravestones galore?
They're gazillions of teeth!

All about the GREAT WHITE SHARK'S
Tooth Machine

The magnificent, muscular great white shark is an apex predator and alpha carnivore – and its jaws are the *most* *powerful* of any shark species.

A shark's smile contains a deadly conveyor belt of 300 teeth, which it uses to grab 13 kilograms of meat in a single bite!

The white shark has several rows of 'spare' teeth on the lower jaw, as well as a main set. These really do act like a conveyor belt, providing them with a constant supply of teeth!

Each tooth contains special nerve cells, which gives them a super touch-sensing ability.

These teeth can rotate and retract, like cat's claws. When the mouth of the white shark is open, her teeth move outward. When her jaw is shut, her teeth move inward.

White sharks can live for 70 years and will get though around 20,000 teeth in a lifetime.

Unsurprisingly, then, the ocean floor is littered with gazillions of discarded shark teeth...

Someone is singing
a strange, lonely tale –
shall we swim for a while with
the song of the whale?

The SONG of the HUMPBACK WHALE

Down in the ocean
and deeper below,
a humpback is gliding
along with the flow...

Solo he journeys,
alone through the sea.
Oh the whale is as lonely
as lonely can be...

26

All over the world
he is famous, renowned,
for his infrasound songs
singing, *"Hey, I'm around!"*

Finding his way between
water and ground,
his song makes a map
in a landscape of sound...

The notes travel quickly,
repeating and changing –
a musical movement
of data exchanging...

"Unbroken, unending,
the songs I am sending
revolve through the seas
and evolve 'til they're trending!"

His sighs are a symphony,
pulsing and shaking,
vibrations that stretch
under waves ever-breaking...

*"Tales of my ancestors
roaming the ocean —
I sing of each one
in a song of devotion..."*

These undulant melodies
ripple and flow
to the heavens above
from the waters below...

"I sing of my journey, I sing of the deep.
Every moment of song is a memory I keep..."

An island in motion through oceans so vast,
where the future is always in tune with the past...

"Every year a new song
I send into the blue,
until all that I sing
is a song just for you."

Under skies full of stars
after touring the Earth,
he arrives once again
at the place of his birth...

"My journey is lonely –
I'm longing to find
that special companion
who's one of a kind..."

And the humpback,
so beautifully blacker than blue,
carries on with his song,
singing *"Love, love me do..."*

But wait! There's a shimmer...
a shadow moves past.
Could it be?
It is she!
He has found her – at last!

All about the HUMPBACK WHALE'S SONG

Humpback whales sing extraordinarily complex songs that can last for hours. Around 30 million years ago, whales evolved these long, low songs to communicate in the darkest depths of the ocean.

Sound travels four times faster underwater than it does through air. These songs of longing can stretch across the horizon into distant oceans.

Their songs are made up of many different noises, forming patterns and sequences, repeating harmonies that make each song unique.

Although both males and females make sounds, it's only the males who sing.
They sing songs to share information between humpback populations...
Songs to map and track their underwater world in landscapes of sound...
Songs to build relationships in the darkness...

Every year, whales migrate across the seas to the place of their birth, singing as they go...

And in the vastness of the ocean, they need to make a big noise.

As only males do the singing, it's thought they must be trying to find a mate.

Not starting or stopping —
our journey is bending,
and folding through moments
of time never-ending...

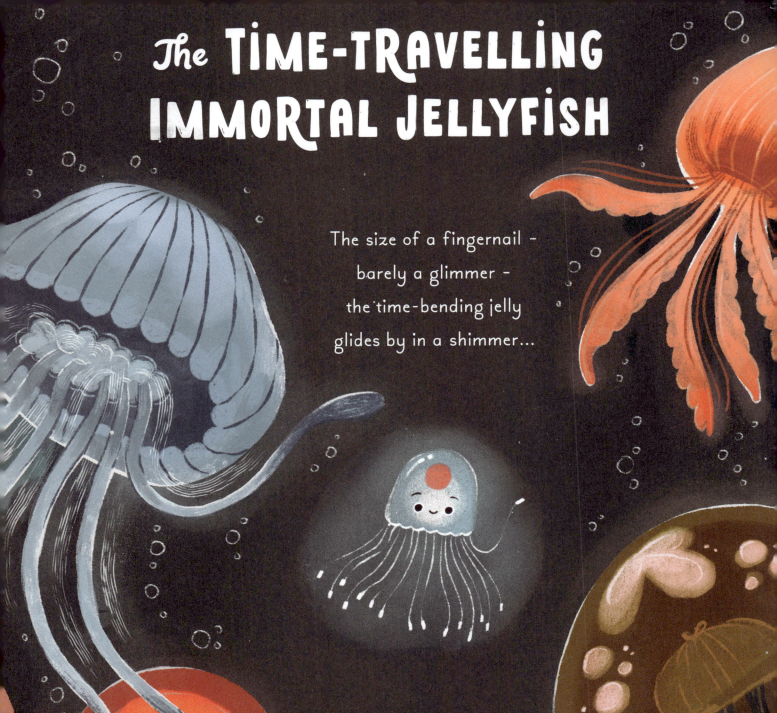

The TIME-TRAVELLING IMMORTAL JELLYFISH

The size of a fingernail –
barely a glimmer –
the time-bending jelly
glides by in a shimmer...

In tropical waters
surrounding the coast,
she floats through the waves
like a miniature ghost...

"My story is long —
it is more than one life,
moving forwards and backwards
through pleasure and strife..."

"Exploring the oceans
with every breath,
I am endless — eternal,
I often cheat death..."

"My name is immortal —
I'm tiny and squishy.
My body is wobbly
and quite jelly-fishy!"

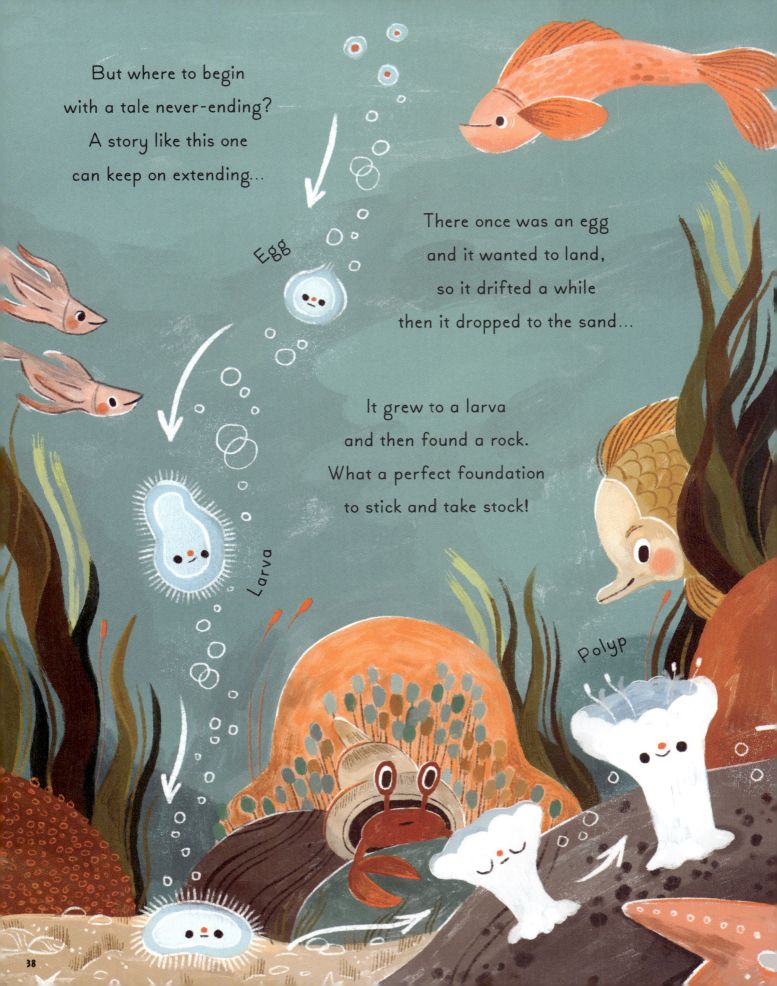

But where to begin
with a tale never-ending?
A story like this one
can keep on extending...

Egg

There once was an egg
and it wanted to land,
so it drifted a while
then it dropped to the sand...

It grew to a larva
and then found a rock.
What a perfect foundation
to stick and take stock!

Larva

Polyp

From larva to polyp,
it didn't stop growing,
until like a tree
tiny branches were showing...

Medusa!

The branches grew buds
and the buds broke away.
It was juvenile now
and it wanted to play...

"My life is for living —
this isn't a game.
Who knows what's in store
when Medusa's your name!"

Wafting and flowing
from ocean to shore,
Medusa keeps going
until she's mature.

With tendrils like threads,
and a glowing red tummy,
she's almost grown-up
and will soon be a mummy...

But WAIT - there's no food.
Then the water goes COLD...
*"I am lost — I am shrinking,
I'll never get old!"*

As the jellyfish falls
to her soft sandy bed,
she begins to get younger
and younger instead...

Her tendrils retract
and her body gets small,
until - oh, she is barely
a jelly at all...

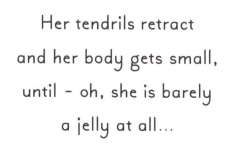

From cyst back to polyp
her life will transform.
(With reverse metamorphosis,
this is the norm...)

"A gateway through time is the way of immortals,
but we tiny jellies don't need to use portals..."

"I'm enduring, undying,
I tend to be friendless.
For this is the way
when your life-span is endless..."

"Yes, I am immortal
and I am a winner...
(unless I turn into
a predator's dinner!)"

All about the IMMORTAL JELLYFISH

The immortal jellyfish is quite possibly Earth's longest-living creature – or at least, has the ability to be. Because this tiny animal can turn back time using something called 'REVERSE METAMORPHOSIS'.

1. A minuscule egg hatches into a larva, called a planula.

2. The planula/larva attaches itself to a rock or hard surface.

Reverse metamorphosis happens if a medusa is injured, starved or drifts into waters too hot or too cold. They drop to the seabed and transform into a ball of tissue called a cyst. This grows into a polyp again and so the jellyfish continues its life-cycle.

3. There it turns into a polyp and grows little branches.

4. Buds break off the polyp, each becoming a juvenile medusa.

5. All being well, these grow into mature medusas. Females will lay eggs and males will fertilise them.

Like fluttering ribbons with curious features,
who knows where they go, these mysterious creatures?

The MIGRATION of the MAGICAL EEL

Our story begins
at the place it will end,
when a shimmer of swimmers
begins to descend...

Every one is a strange
and mysterious fish:
every one is an eel
that has swum with one wish...

To return to this ocean
of indigo blue,
just in time for
a marvellous eel rendezvous!

Every eel has arrived for one reason: to breed
in the orangey bloom of the Sargassum weed.

A massive migration
all coming to spawn.
Now from millions of eggs
many millions are born!

So the babies are off!
On the current they float
through the ebb and the flow
every one like a boat.

As soon as they hatch
with a sigh of relief,
every one of them looks
like a miniature leaf...

The larvae declare,
"We would so love to stay
but there's no time to lose —
we must be on our way!"

Their journey continues –
who knows what's in store,
or the dangers they'll meet
before reaching the shore...

And something is happening...
See? As they pass
every eel has become
as transparent as glass...

Tiny spaghetti strands
swimming with skill,
on and on they keep swimming
and swimming until...

49

They turn into elvers
that wriggle and swerve
with the flick of a snake
in a serpentine curve...

As slender as willows
they ripple and fly
into rivers and streams
under watery sky...

So, upstream they surge
like a flutter of tape,
each eel is emerging
in freshwater shape...

"These are my rivers
and here I belong.
I will stay and keep swimming
until I grow strong!"

Cylindrical body
all silky with slime,
the eel can grow up
to a metre in time.

From yellow to silver
with slip-slimy skin,
made for riding and gliding
and slip-sliding in!

For years and for years
in the rivers they stay,
'til the moment arrives
and the eels must away...

They ripple like ribbons...
They billow and sweep
towards oceans of water
more salty and deep...

They go with the flow
and they swim with the tide,
these mysterious fish
on their magical ride!

Returning once more
to the place they were born...
They have made it – amazing!
And now, time to spawn...

All about the MAGICAL EEL'S
Epic Migration

Mature eels return to the Sargasso Sea in the North Atlantic Ocean every year to spawn, the same place they were born some 40 years before...

2. They hatch and the tiny, leaf-like eel larvae continue on their epic journey of over 6,000 miles!

3. It's only when they reach the European continental shelf that they change into an eel-like shape. Now they look like transparent spaghetti!

1. Millions of minuscule eel eggs are carried along on warm currents.

Mysterious, snake-like and sometimes alarming, the eel is disarming but ever-so-charming...

6. After many more years of swimming and eating, the eels reach maturity. These are 'silver eels' and when the time is right, they ready themselves to return to the wide Sargasso Sea.

5. The juniors mature and their skin changes again. They are now 'yellow eels'.

4. On arriving in freshwater rivers, the eels' skin darkens and they become 'elvers'.

Here is a forest. Let's wander inside
to the world of a cephalopod –
where does she hide?

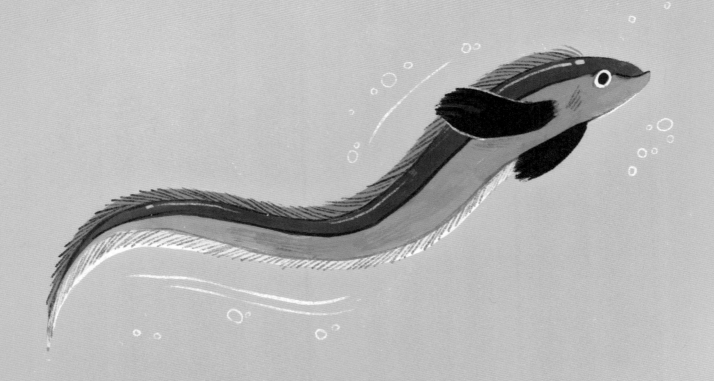

The SHAPE-SHIFTING OCTOPUS

You'll find her in filigree
shadows that sway,
where the cuttlefish scoot
and the pufferfish play...

Sensational, sensitive,
super at healing:
her body is soft – it's a body for feeling!

A shape-shifting drifter
though whispering night,
she has skin like a screen
made of shadows and light...

"Here is my home
in this garden of green.
It's the loveliest garden
you ever have seen!"

"Slippy and supple,
the wonderful weed
is a sheltering wood
holding all that I need..."

The octopus moves
like a parachute blowing,
ballooning and billowing –
where is she going?

Swooping round coral
and over some shells
with a glittering shimmer
of silvery bells.

Her skin becomes textured,
her body transforms –
what magical camouflage tricks she performs!

And here she will wait for some good things to eat.
Maybe clams or a lobster – her favourite treat!

But what was that flickering – is she alone?
Or is somebody hiding behind the large stone...?

OUT sweeps a shark from a gap in the rock,
and he smiles: *"Oh, hello... did I give you a shock?"*

"YES!" cries the octopus,
waving her arms
as she scoots through the shoots
and away from his charms.

Quickly she flies, crying,
"Help me... oh HELP!"
as she weaves through
the forest of towering kelp.

Wafting and zooming
round foliage blooming,
the octopus races,
her predator looming...

No bones in her body,
no bones in her head,
but brains – she has many,
"I'll use those instead…"

And suddenly squirting
the inkiest black,
she is lost in a cloud
and the shark looses track…

She races ahead through the emerald flow,
while the shark lags behind thinking, *"Where did she go…?!"*

With a flash, then a glimmer
the water is clear.
"So long!" She is gone
and will soon disappear…

Then spotting a gap
in the crack of a rock,
she manoeuvres inside
where she stops to take stock.

"No bones in my body,
no bones make me free...
But brains — I have many —
I'm clever, you see!"

"Sensational, sensitive,
super at healing:
my body is soft —
it's a body for feeling!"

A shape-shifting drifter
through whispering night,
she has skin like a screen
made of shadows and light.

All about the SHAPE-SHIFTING OCTOPUS

Octopuses have one of the best shape-shifting camouflage abilities in the world. They are seriously clever, having not one but NINE brains - one big 'main' brain, plus a mini-brain in each of their eight arms.

Every arm can touch, taste and move independently. And special sensors in their suckers mean they never get tangled up!

An octopus's brain-to-body ratio is the biggest of any invertebrate. They are brilliant at solving puzzles and performing tasks. They can even use tools to build homes!

Having a soft, flexible and entirely boneless body means that they have complete flexibility. Extending, bending, contracting and flexing in any direction... the possibilities are endless!

A boneless body means they can squeeze into tiny nooks, crannies and spaces. They can even regrow a lost limb!

The ocean is stormy –
a north wind is blowing...
Antarctica beckons
and that's where we're going!

A PENGUIN'S SQUEEZE
in the FREEZE

Two emperor penguins
were happy to meet
in the icy Antarctic
all covered in sleet...

There weren't many places
the penguins could go,
so they swam in the ocean
and skated on snow.

A beautiful egg
made their family complete.
And then Ma said to Pa,
"Keep it warm on your feet!"

"Farewell now!" she whistled,
"I know it seems rude
but I need to get going
to find us some food..."

"I won't be too long,
just be strong — and goodbye!"
She said, "Keep our egg cosy..."
Said Papa, "I'll try!"

She slid on her tummy
and dived in the sea,
calling, "Watch out, you fishes,
it's time for my tea!"

The emperor papas
have all stayed behind.
"It's our job!" they declare,
"to look after our kind!"

Slowly rotating,
they cuddle up tight
through the dark winter days
and the long winter night...

After 65 days...

every egg in its pouch
is beginning to hatch...

with a crack then an,
"OUCH!"

It's chilly outside,
but no matter - each chick
is equipped with a coat
that is fluffy and thick.

As the days turn to weeks
since their mothers' goodbye,
"Will she ever come back?"
all the little chicks cry.

The sea-ice is shifting...
oh no - are they drifting?
Then a swell of emotion
erupts from the ocean!

As thousands of penguins
begin to arrive...
squawking, *"Mama is coming!"*
"I'm back – I'm alive!"

But how will each mother
find each little baby?
In thousands of penguins –
Impossible!
Maybe...

For emperor penguins
can all make a sound
that's completely unique.
So each baby is found!

After months on the ice
and a very long fast,
the papa hands over
to Mama - at last!

"I kept our egg safe
with the help of my feet,
and now it's MY turn
to find something to eat!"

All about the EMPEROR PENGUIN

Emperors are the largest penguin species, standing at around 1.2 metres (4 feet) when fully grown. They live in Antarctica on great sheets of sea-ice, and dive deep in the icy ocean to hunt for food.

Between May and June, the female lays a single egg and carefully passes it over to the top of her partner's feet. Then she leaves to find food, while her mate stays behind to care for their egg.

This he does for an incredible nine weeks, sometimes more, keeping the egg above the freezing ground on his feet, covered by a flap of skin called a brood pouch.

As soon as the females return - plump and well fed - they find their partner by using a unique call. Once reunited, they regurgitate food into their chick's hungry mouth. And finally, the papa penguins head off to find something to eat!

After around 65 days of incubation, the eggs hatch and the fluffy penguin chicks emerge.

Colonies of around 5,000 males huddle together for warmth. To keep things fair, they slowly rotate, shuffling around to take it in turns to be in the warmer centre.

In temperate waters,
not far from the shore,
a shy little creature is
roaming the floor...

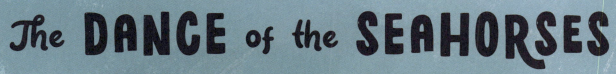

The DANCE of the SEAHORSES

"I am delicate, small
and my name — why of course —
comes from having a head
like a miniature horse!"

Although he is slow,
he's an upstanding male,
with a super-long neck
and a beautiful tail!

He swishes along
then he clings to the weeds,
where his two fishy eyes
will detect what he needs.

He says...
"All bony and spiny, I'm ever-so tiny,
I wait for the love of my life.
A nozzle-nosed prancer,
a wonderful dancer...
she's all that I want from a wife!"

She says...
"All bony and spiny, I'm ever-so tiny,
I flicker and flit through the sea.
I don't want a chancer — no!
Only a dancer,
to do all his dancing for me!"

"Yippee!" he declares.
"Right on time — never late!"
As their two tails entwine,
"Shall we go on a...date?"

And twirling together
they shimmy and prance
for the rest of the day
in a wonderful dance...

From the tips of their tails
to the nibs of their noses,
the seahorses sway,
striking several poses...

They spin and encircle,
their two tummies touch,
then she says, *"Take my eggs —*
for I love you so much..."

The male has a ready-made
pouch for their brood.
It is here he will keep them
and give them their food.

Off she slips through the sea
with a bob and a swish,
to go hunting herself
for crustaceans and fish.

Although she has gone
leaving papa alone,
every day she returns
so he's not on his own...

"*Good morning!*" she bobs,
and the little fish kiss,
then continue their dancing
in marital bliss.

"How are the babies?"
she asks, "Are they fine?"
And her partner replies,
"They are simply divine!"

While their parents embrace
as they dance to and fro,
safe and snug in his tummy
the sea-babies grow.

And 40 days later
it's time to deliver:
the babies are born
with a whoosh and a quiver!

Ever-so tiny they flick
and they prance,
While "Goodbye!", cry their parents,
"Now on with the dance!"

All about THE SEAHORSE

Seahorses are shy, enchanting little fish, named after the fact that their head looks like a horse. They have long, bony bodies and swim upright, using their fins to steer.

There are around 40 species of seahorse and these tiny, spiny predators live in the calm, weedy shallows of warm, temperate seas.

Seahorses pair up in a partnership that lasts for life. Their first date is an intricate dance that can last for up to 8 hours.

Once a partnership has been established, the female seahorse transfers her eggs into her partner's pouch. This is specially designed to work like a womb, protecting, feeding and removing any waste from the growing eggs.

Then the female leaves to feed on shrimp and fish eggs. Each day she returns to her partner and together they dance around their ocean home together.

After between 30-45 days, the seahorse daddy is ready to give birth. Strong contractions spray showers of baby seahorses into the water. These minscule 'fry' cling together as they are carried away by the current.

Their parents continue to dance together every day in a tender, delicate seahorse ballet, and soon many more babies will be on the way!

The Moon rises high,
gleaming silver and ghostly.
The beach is deserted
and quiet – well, mostly...

The SALTY TEARS of the SEA TURTLE

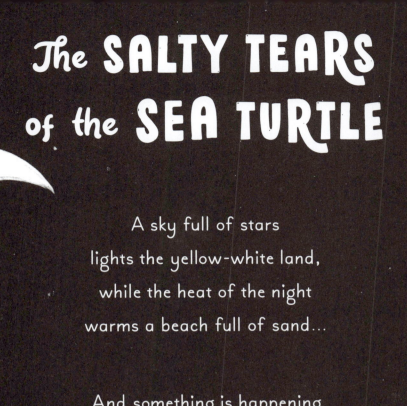

A sky full of stars
lights the yellow-white land,
while the heat of the night
warms a beach full of sand...

And something is happening
under the ground -
tiny voices are calling
without any sound...

Soon there's a scooping,
a digging and scratching...
The sand is expanding
and something is hatching!

Flotillas of miniature
turtles emerge
and immediately move
to the sea in a surge...

They shuffle and scrape,
in a scuffle they hurtle –
a moving and magical
carpet of turtle!

Scooting en masse
to the edge of the sea,
singing, "We may be mini
but many are we!"

"We hatch all at once
in a massive arrival.
The more of us upping
our chance of survival!"

"And though we are helpless
and terribly small,
we crawl to the ocean,
come one and come all!"

Into the water, now
bobbing and floating
with flippers a-flapping
as if they are boating.

Their life has begun with goodbyes,
and now friendless,
the turtles set off on
a journey that's endless.

The pull of the tides
tosses each one around,
and a few disappear,
never more to be found...

But some keep on swimming -
they're fearless and plucky,
*"But mostly we know
we're incredibly lucky!"*

"We swim through the ocean,
keep swimming until
our shell grows enormous
along with our bill!"

As the days turn to weeks
and the weeks become years,
every turtle, each one,
sheds an ocean of tears.

"Don't worry," they say,
"We're not sad like you think.
We're just shedding the salt
from the water we drink!"

The Earth is a magnet,
the forces she yields
are some powerful lines
known as a 'magnetic fields'.

These fields guide the turtles
right back to the beach
where each one of them hatched –
it's the place they must reach!

Here comes a mightily,
tired-out turtle!
She sweeps through the surf
with a last-minute hurtle...

She crawls up the beach
and she finds a good patch,
knowing this is the place
where her babies will hatch...

Her eggs safely buried,
she crosses the shore
and returns to the ocean
to journey once more...

*"My internal compass
is feeling the force...
I need to keep swimming
and swimming – of course!"*

The moon rises up
looking pallid and ghostly,
the beach is deserted
and quiet – well, mostly...

All about the GREEN SEA TURTLE

The green sea turtle gets its name from the green fat found beneath its shell. Turtle tears come not from sadness, but salt excretion glands in the corners of their eyes, which 'cry' tears to get rid of the excess salt in their body.

1. The baby turtles hatch and emerge from the nest to find the bright horizon.

2. As soon as they've set eyes on the light of the ocean, the hatchlings set out to sea, following the sound of the waves.

Green turtles have an excellent internal compass but can still get lost or confused by artificial lights or noises.

5. Female turtles carve out an egg chamber about 75 centimetres deep in the sand, where they lay their eggs. The eggs incubate for 49-62 days before hatching.

3. Once they reach maturity, females will return to the beach where they were born.

4. Each beach has its own, unique magnetic pattern and this pattern is imprinted in the turtle's brain. So they navigate their way home, guided by the Earth's magnetic fields.

Astonishing creatures
and curious fish,
every one is a wonder
with only one wish...

From corals to crevices,
even each shell
has a wonderful
5 minute story to tell!

FURTHER READING

Oceanarium by Teagan White and Loveday Trinick

The Big Book of Blue by Yuval Zommer

Earth's Acquarium by Alexander Kaufman and Mariana Rodrigues

Explore every bit of this book with a wondrous eye!
– Mona K

Inspired by the work of conservationists and photographers Paul Nicklen and Christina Mittermeier, and the charity Sea Change Project.
– Gabby D

MAGIC CAT PUBLISHING

5 Minute Ocean Stories © 2024 Lucky Cat Publishing Ltd
Text © 2024 Gabby Dawnay
Illustrations © 2024 Mona K
First Published in 2024 by Magic Cat Publishing, an imprint of Lucky Cat Publishing Ltd,
Unit 2 Empress Works, 24 Grove Passage, London E2 9FQ, UK

A catalogue record for this book is available from the British Library.

ISBN 978-1-915569-30-1

The illustrations were created using gouache, coloured pencil and digital media
Set in Bakerie, Claytonia, Pistachio and Quicksand

Published by Rachel Williams and Jenny Broom
Designed by Nicola Price and Maisy Ruffels

Manufactured in China

9 8 7 6 5 4 3 2 1